SAVE OUR PLANET
"GO GREEN"

REDUCE, REUSE, RECYCLE, RENEW

Sandi Pope Pres., Founder
CISTA 4 CLEAN WATER INC.
None-profit, 501c3 Environmental
Corporation

CISTA 4 CLEAN WATER INC. is a not-for-profit, 501c3 Environmental Organization committed to heighten the awareness of the polluted state of our oceans and waterways, endangered eco system, and the diminishing life on our planet, due to Global warming.

Our Global mission, CISTA GIRLS leading the way to save our planet. Since our inception in 2016, Girls, 8-18 years old, have become members of the CISTA GIRLS Environmental Club, to   volunteer and participate in environmental projects, such as beach clean-ups, science fairs, host TV shows, spelling bees, recycle projects, library presentations, and more, to educate the community on how to build a stronger relationship between man and the environment.

The CISTA GIRLS ENVIRONMENTAL CLUB is more than a club. Members are provided FREE TRAINING for projects and programs that will enhance their personal and professional life's achievements. They have acquire skills, as leaders, writers, public speakers, organizers, television hosts and most important the value of community service. View CISTA GIRLS History at a Glance: www.cistagirlmagazinecom.   Contact information: www.cistagirlveenus.com, cistagirlveenus@gmail.com.

TABLE OF CONTENTS

"GO GREEN"

INTRODUCTION TO "GO GREEN"

GO GREEN means being more environmentally aware and changing your behavior and lifestyle to reduce the amount of pollution and waste you generate. Any action you take that contributes to sustainable living makes a positive impact on the environment and global warming. Going Green can include a few simple but powerful ideas.

1. Pursue knowledge and practices that lead to more environmentally friendly decisions.
2. Use environmentally-friendly products.
3. Protect the environment and sustaining its natural resources.
4. Reduce air pollution and environmental toxins.
5. Reduce the need for paper go digital.
6. Reduce energy consumption. Turn lights off when you leave a room.
7. Weather-strip your home to make it more energy-efficient.
8. Air dry your clothes to save energy.
9. Installing a rain barrel to save water.
10. Recycling your pillows, jeans, socks, or e-waste properly.

GO GREEN AND ECO- FRIENDLY

- **Eco-friendly** is a term typically used to describe a product and how it's made. For a product to be genuinely eco-friendly it has to take both earth and human health into account. The resources used to make the item need to come from sustainable resources. For clothing that means the materials used are grown without the use of harmful pesticides and herbicides. Organic wool, hemp, and natural cotton are a few examples.

- The product end of life is made with materials that can be either reused or recycled like glass, wood, metal, or cardboard. Or is it made from biodegradable material that can be composted which helps to reduce the amount of waste heading to the landfill.
- Eco-friendly cleaning & beauty products are made from non-toxic, natural ingredients that don't poison us or the planet and are sustainably sourced and grown.
- It also means that the company that is making the product you are buying is also taking all of the above into account.

HOW DOES GOING GREEN IMPACT OUR BODY AND MIND?

There are so many benefits to living a more mindful life. Studies have shown that eating a more plant-based diet can help both the planet and your health. Being more green means spending more time in nature, hiking, walking, biking or exercising. This adventure can even help you sleep better. Spending at least 30 minutes a day in nature can be very beneficial to your mental stability.

This type of lifestyle also helps to reduce toxins from common things like cleaning products and beauty products. Both of these can have harmful effects on your health. Conventional cleaners give off VOCs (volatile organic compounds) which create indoor air pollution and can cause asthma and other respiratory issues. Beauty products that contain synthetic ingredients like phthalates have also been shown to be quite harmful to our health.

THE BENEFITS OF GOING GREEN

Global warming is a serious issue. Our planet is a beautiful, awe-inspiring place filled with magnificent ecosystems and biodiversity that is under immediate threat. We have melting ice and rising sea levels. Massive wildfires, animal and plant extinction, deforestation, and plastic pollution just to name a few. Never before has it been more pressing than now for humanity to step up and make the changes needed to keep our planet and humans safe and alive.

Going green is about getting back to basics and choosing to simplify life. It's about making new choices, ones that are healthier for you, the planet, and the next generation.

Benefit

1. Reduce Pollution

When you make mindful decisions based on the impact on the planet, you are having a direct effect on reducing soil, water, and air pollution. Choosing to compost your food for example reduces methane gas produced in landfills which will, in turn, lowers greenhouse gas emissions. If you choose organic food you are helping to keep harmful pesticides out of our soil and water. Recycling and reusing means less natural materials are being taken from the earth to make new products.

Benefit

2. Preserve Wildlife

Ecosystem experts estimates an extinction rate of up to 8,700 species a year. The paper you use daily is responsible for massive deforestation, and the same goes for your toilet paper. By choosing recycled paper products you are aiding in slowing the rate of extinction for animals living in the forest. Not only are woodland creatures at risk, but so is marine life. Our oceans and lakes are filled with all kinds of pollutants, one of the biggest threats being plastic.

Benefit

3. Reduce Consumption

Reducing consumption means creating less waste (especially in the kitchen) which means fewer things you have to toss in the garbage which means less of it has to go to landfills. The food we eat, the energy we burn, and the things we buy are all tied to greenhouse gas emissions. Reducing consumption helps to limit greenhouse gas emissions associated with the production of goods and services. Green living is about moderation, efficiency, and living less expensively. The more stuff you have the more you have to spend to take care of it.

Benefit

4. Save Money

If you are choosing to get rid of paper towels, you can either invest in un-paper towels or you can simply cut up old towels and t-shirts and use them as rags. The same goes for napkins, why are you buying paper ones over and over again when you can use cloth? The average family spends over $800 annually on paper towels and paper napkins. These smaller actions can save you a ton of money in the long run. Turning off the light when you leave the room save you money on your electric bill.

Benefit

5. Bring Home Less Plastic

Around one-fifth of all households waste, comes from food packaging. Always make sure you have enough reusable bags with you when you shop. Choose foods that come unwrapped (naked) or in glass, cardboard, or aluminum.

Shop at local bakeries and butchers for fresh items that come in paper bags or paper wrapping. Ask not to include disposable plastic trays, and plastic cutlery when you order take-out. Avoid plastic razors (try a safety one instead) plastic toothbrushes (go for bamboo) and cleaning products that come in plastic.

Many cleaning products and cosmetics are now produced in refillable containers and greener shops are popping up all over offering refills for these types of products. Dried goods like rice, cereals, and coffee can also be bought in bulk with little to zero packaging.

Packaging lotions, shampoos and other types of cosmetic items are very heavy and transporting them uses a lot of fossil fuel. Conventional wipes are made from plastic (polyester) and create fatbergs when they are flushed. (Never flush wipes). Fatbergs are large masses of solidified fat that when flushed it coagulates into the sewer.

Benefit
6. Choose eco-friendly cleaners that have less toxic ingredients.

Benefit
7. Avoid fabric softener
Fabric softeners can contain harmful ingredients, including:
Benzyl acetate: Linked to pancreatic cancer
Benzyl alcohol: Linked to upper respiratory tract irritation
Methylisothiazolinone: A potent skin allergen
Glutaral: Known to trigger asthma and skin allergies
D&C violet 2: Linked to cancer
Some fabric softeners also emit chemicals that can cause adverse reactions in humans. These chemicals can include: Formaldehyde, PFAS.
 (add a cup of vinegar during the rinse cycle) and dry sheets, (use wool dryer balls) they reduce drying time energy.

When buying new clothes try to buy ones that are made from natural fibers, like hemp, organic cotton, bamboo, jute, or linen. Synthetic clothing like elastane/spandex or nylon creates microplastic that breaks off in the wash and heads through the sewage system right into our lakes and oceans. Try reducing how often you wash or you can try the Guppy Bag, it helps capture some of that plastic.

9. Make Your Home More Efficient

Our homes produce about 30 percent of the emissions that contribute to climate change. With a few simple energy-saving adjustments you can reduce your energy consumption.

About three-quarters of the energy we use to heat and cool our homes is lost due to poor insulation of windows and doors. Weather-stripping your home is one of the most effective things you can do, in addition, to draught-proofing. Look for insulation made from wool, hemp, or flax.

Heating your home accounts for more than half of the energy used by the average home. An easy and effective way to cut down costs is to turn down the heat. Adjusting your thermostat by one degree can save you about 10% on your heating bill. If you are renovating your home look for low-carbon heating and hot water systems.

Choosing the most efficient household appliances will also make a big difference in your energy bill. Choose energy star-certified appliances. Newer plasma TVs are more efficient than older ones. LED TVs are better. Laptops and desktops use the same amount of energy. Best to look for devices that have a high-energy rating and low standby power consumption. A good energy saver try to remember to turn off and unplug.

So many things don't need to be tossed, and can be fixed. Repairing and refurbishing is one the best ways to go green. Choose products that are designed to last longer.

10. Change how you Cook and Eat

The food we eat and how it ends up on our plates takes up a lot of resources. Water use, food miles, food waste, food packaging, and unsustainable farming practices all play their important part.
Make a list before you shop. Only buy what you need and shop in bulk. Buy food that has a lower carbon footprint, that means reducing meat consumption.

Meat production and the methane gas created from the production makes it one of the biggest contributors to global warming. Farmed animals like cows and sheep are responsible for nearly 30 percent of the total methane from meat consumption.

Industrial farming of beef uses 28 times more land and 11 times more water than any other industry. It's important to try Meatless Mondays and a plant based diet. Eating local and organic reduces food miles substantially. About 10 percent of the total carbon footprint of your food comes from the distance it has to travel. Connect with local farmers, support farmers markets and Community shared gardens.

11. Get Rid of Stuff

Green living promotes less consumption and MORE recycle.
Use your community recycle Bin. It's local. It's Free.

START GREENING
BUILD A BACK YARD
GREENHOUSE

Green your backyard. Make use of your
plastic bottles. Build a green house.

SIMPLE STEPS TO BUILD YOUR BACKYARD GREEN HOUSE

1. Remove and recycle the lids.

2. Wash and remove the labels from the bottles

3. Cut off the bottom of each bottle, can use scissors.

4. Size: 6' long walls facing two 8' long walls. 7' high.

5. Fix four lumber posts into the ground, to be the corners of the greenhouse.

Use recycled lumber.

6. Build frames for three of your walls, short wall will contain your door.

7. Use nails or screws to attach the plastic bottles to the frames.

8. Cover and secure the hothouse with a piece of plastic sheeting. This will help to keep the heat in and the cold out.

9. Enjoy your fresh fruits and vegetables

10. ON LINE provides more details

PART
A
"REDUCE"

BUY LOCAL PRODUCE TO REDUCE TRAVEL TIME AND EMISSIONS.

PART A
"REDUCE"

CHAPTER I
REDUCE EMISSIONS

Reduce Food Mileage helps the environment.

One of the most important ways buying locally helps the environment is by reducing the travel mileage transporting food. By shopping locally, you can purchase fresh produce right from your community. Many of the food items you buy travel over 1500 miles or more to reach where you live. By cutting down on these miles, you are reducing the environmental impact of the food you eat. Local food doesn't create large carbon footprints from overseas shipping and airplane transportation. This cuts down on fuel consumption and air pollution.

Food products are more accessible.

It's easy for local businesses to bring their products to their consumers because their consumers are nearby. At a local farmers market consumers are able to easily access lots of local homegrown produce, without leaving their own community. Shoppers are able to easily walk or bike to their local farmers market.

Fresher Produce

Consumers are able to enjoy produce that is fresh and nutritious, organic, hormone free and pesticide free. Keeping harmful toxins, like pesticides, out of the air helps to improve crops and air quality. In addition, because the produce is fresh and brought directly from farm to the table, there is less waste. Many large retailers have significant food waste due to items going bad before they are bought, before they are ripe, which creates trash and garbage problems and environmental hazards.

Protects Local Land & Wildlife

Buying local also helps to protect local lands and wildlife. By buying local, you are supporting local farmers and producers. With your support, these farms are able to stay in operation. Because the farms are owned and operated by local farmers and producers, they aren't being sold to local developers. Local developers could completely transform the land, devastating the wildlife. Big business producers could buy out the farm and incorporate inhumane and non-eco-friendly farming practices.

Local Workforce

The great environmental benefit of buying locally supports the local workforce. Buying your groceries at the local farmers market, will be helping to keep local growers, creators and farmers in their jobs. You're also creating an opportunity for other local jobs such as the team who organizes the farmers market, the team that sets up the stalls, the team that cleans up at the end of the day, etc. All of these local businesses with local workers are in place because consumers are appreciating and supporting available fresh produce.

COSTLY TRAVEL TIME AND EMISSIONS TO GET FOOD TO YOUR TABLE.

CHAPTER II
REDUCE PLASTIC OCEAN DUMPING

Our ocean is in serious trouble. Heating, pollution, acidification, and oxygen loss pose serious threats to the health of the ocean and to all Aquatic life. Our oceans takes care of us in so many ways.

1. The ocean regulates our climate and provides the air we breathe

Our ocean absorbs 25 per cent of all carbon emission, while generating 50 per cent of the oxygen we need to survive. It not only functions as the lungs of the planet, providing us with the air we breathe, but also as the world's largest carbon sink helping to combat the negative impacts of climate change. Additionally, the ocean has taken up more than 90 per cent of the excess heat in the climate system helping to regulate temperatures on land. Thus, climate action depends on a healthy ocean, and a healthy ocean requires urgent climate action.

2. The ocean feeds us

The ocean and its biodiversity provide our global community with 15 per cent of the animal protein we eat. In less developed countries, seafood is the primary source of protein to over 50 per cent of the population. It is therefore critical to protect the ocean's biodiversity and practice sustainable fishing strategies for continued consumption. Currently, more than 10 million tons of fish go to waste every year because of destructive fishing practices.

3. Our oceans provides jobs and livelihoods

The ocean provides livelihoods to 3 billion people, nearly 50 per cent of the entire global population. Marine fisheries provide 57 million jobs globally. The blue economy is a strong industry that allows many to make their living and provide for their families. However, over 60 per cent of the world's major marine ecosystems that supports these livelihoods are being used unsustainably, with a significant portion being completely degraded. Additionally, research states that pollution from 11 million tons of plastic that enters the ocean annually, costs an estimated US $13 billion, including clean-up costs and financial losses from fisheries and additional ocean-based industries.

4. The ocean is a tool for economic development

The ocean is a significant economic tool. Ocean economies are among the most rapidly growing in the world. The market value of marine and coastal resources and the developing industry is estimated to be US $3 trillion per year, which is about 5 per cent of total global gross domestic product. Thus, developing countries' access to the ocean and shorelines allow them to develop and attract foreign direct investments and direct industry production within the state. Additionally, 80 per cent of tourism happens in coastal areas. The ocean-related tourism industry grows an estimated US $134 billion every year. However, for states to utilize their ocean resources, we must work together as a global community to protect the ocean. It is estimated that the loss of tourism due to coral bleaching alone is as much as $12 billion annually. With ocean levels rising as the temperature of our planet increases, coastline-specific tourism and energy industries are at risk along with the 680 million people who live in low-lying coastal areas.

5. Our oceans must be healthy to survive

The ocean affects all of us. It provides climate regulation, food, jobs, livelihoods, and economic progress. For the future survival of our planet we must pay attention and reduce climate change conditions.

CHAPTER III
REDUCE SINGLE USE PLASTICS

Not only plastic water bottles, but all other things of plastic. According to the World Economic Forum, just 14% of plastic packaging is collected for recycling globally. And because of complexities in the recycling process, huge amounts of single-use plastic (as well as glass and cardboard) that consumers try to recycle ultimately end up getting burned or tossed into landfills anyway. If recyclable materials are contaminated by food waste, or if consumers misunderstand what can be recycled, their garbage may not end up being repurposed. A 2017 study estimated that, of all the plastic waste generated globally just 9% had been recycled, while 12% was incinerated and the rest ended up in landfills.

There are arguments that instead of the recycling process, companies should replace single-use containers with those that can be used over and over again, perhaps a durable metal or glass container that can be refilled either in a store, or in a consumers' home. Using the same containers, in the same form, over and over again ideally eases demand for virgin materials, reduces energy needed to produce thousands of new plastic bottles or cardboard boxes, and prevents heaps of trash from ending up in landfills or oceans.

Some global corporations have piloted reusable packaging programs within the last few years. They have set up refill stations in some popular Mall locations that offers refillable versions of some Beauty and Planet-branded hair products. Some companies deliver laundry detergent and cleaning product refills to customers' homes by electric tricycle. Some companies design their programs around selling concentrated detergents and cleaning products that people can mix with water in reusable containers at home. Surveys also recognized that for the right product category and the right supply chain, the effect of reusable packaging can be a huge benefit to the environment.

Concerns of this program illustrated that if someone has to mail back an empty metal bottle for refilling, there will be an environmental burden associated with transporting it back and forth and cleaning it. A direct to consumer service that gives people all the tools they need to refill at home, like shampoo capsules or detergent concentrates. are a good option because they don't require return shipping and are convenient.

Companies are considering making it "more of an experience for consumers with well-designed refill stations or discounts on repeat purchases. One company is selling about 65% of its permanent products, without packaging. In these cases, the brand sells solids like bath bombs and shampoo bars loose. Liquid products that require packaging, like lotions and shower gels, are sold in pots made from post-consumer-recycled plastic. If customers return five pots to the store, they get a free face mask.

Companies involved with this program states that even with the promise of free stuff, there is only about a 17% return rate. And because it's so difficult to adequately clean the returned pots, they break them down and uses them to make new ones. Doing so reduces the brand's need to buy recycled plastic from other sources.

Companies survey also experienced a return rate of more than 80% on their refillable programs, largely because of the financial incentives it offers: customers must pay a deposit of up to $10 (for a pack of cleaning wipes) for each container they purchase, and it's only refunded when they return it for refilling.

In Summary, the planet needs all the ideas it can get to reduce the effects of Global Warming. Single use plastics and all other plastics continue to diminish the health and welfare of humans and that of the planet as well. We must make a commitment to "RENEW" our thinking on how to help resolve this ongoing problem.

Your ideas on how to reduce single use plastics. How do we remove balloons, an essential product in our birthday parties?

____
____
____
____
____
____
____
____
______________________________.

SINGLE USE PLASTICS

Cutlery, Straws, Shopping bags, Water bottles, Styrofoam, Take-out containers are single use products.

Single-use plastics are products that are designed to be used only once before being discarded. Single-use plastics are made from fossil fuel–based chemicals and are often disposed of right after use. They don't biodegrade, but instead break down into microplastics that pollute our water sources and food.

Reduce single-use plastics:
Use a non-plastic toothbrush, one made of bamboo.
Purchase post-consumer recycled content trash bags.
Use containers with lids instead of plastic wrap to store food.
Use decorations other than balloons.

CHAPTER IV
REDUCE MEAT CONSUMPTION

Meatless Mondays

Meatless Monday encourages people to reduce their meat consumption for their personal health and the health of our planet. Meatless Monday is of utmost importance especially in the United States, as we consume much more animal products than the rest of the world.

The meat industry uses vast amounts of fossil fuels and water and lots of grain to feed livestock, which is extremely inefficient. About 1,850 gallons of water is needed to produce a singular pound of beef, comparable to only 39 gallons of water per pound of vegetables. A vegetarian diet alone could dramatically reduce water consumption by 58% per person!

Meat production also is a major contributor to greenhouse gas emissions, which has proven to correlate to the climate change crisis. Some benefits of eating plant-based once a week include:

Save 133 gallons of water with each meatless meal!

With your participation in a Meatless Monday you could reduce your carbon footprint by 8 pounds each. If you commit to participating every Monday, that would be equivalent to skipping one serving of beef for a year, and saving the amount of emissions produced driving 348 miles in a car.

Benefits of personal health including a Meatless Monday program into your kitchen.

You will Reduce your risk of heart disease and stroke because fruits and vegetables help your body fight cardiovascular disease, obesity and maintaining a healthy weight.

Eating red meat increases the risk of colorectal cancer.

Adding more greens to your diet helps aid in maintaining a healthy weight which is a key factor of preventing type 2 diabetes.

At the end of the year you can calculate enormous savings on your grocery bill, when you go meatless once a week.

EAT MORE VEGETABLES

HOW PLANT-BASED FOODS REDUCE ENVIRONMENTAL IMPACT:

The agriculture industry is responsible for 25 percent of the world's greenhouse gas emissions. Animal farming accounts for three-quarters of those effects. It creates twice as much greenhouse gas as emissions from cars, and the same amount as emissions caused by heating and electricity.

Livestock naturally produce methane as part of their normal digestive processes. (cows passing gas) But methane is harmful because it heats the earth 20 times faster than carbon dioxide, which means it's a big contributor to global warming. When you consider that 32 million cows alone are farmed each year, you're talking massive amounts of methane generated from their collective waste — 300 million tons of it a year accounts for 37 percent of all agricultural greenhouse gas emissions. Eating more plant-based food can help offset the emission production. For example tofu produces just under 7 pounds of greenhouse gases for every 100g of protein, while beef produces 231 pounds for every 100g of protein, that's more than 30 times the impact.

The atmosphere isn't the only thing you're helping by eating more plants. Producing a single pound of animal protein requires 100 times more water than producing a pound of grain protein. And producing one gallon of milk requires 1,000 gallons of water. On the whole, animal agriculture uses up over half of the world's fresh water supply.

Eating plant-based food benefits wildlife and animal habitats, too. Farmland currently accounts for 60 percent of the earth's usable land. Millions of acres are deforested each year to meet the growing demand for cattle, pigs, chickens, and other poultry, and to grow the food that feeds them. This creates serious problems. Naturally occurring plant life is destroyed, and animal species are left displaced with loss of their habitats. Agriculture is the biggest contributor to wildlife extinction.

It also equates to a large amount of waste. Most farms are not used to grow fruits, vegetables, or grains for human consumption, but instead to grow feed for the livestock we eventually eat. That's even more water and resources being used up, and more greenhouse gasses produced.

Conscious consumption is advocated. You need to try to eliminate meat entirely to make a major impact on global warming and climate change. Start by choosing to eat plant-based more often.

EASY WAYS TO EAT MORE PLANTS
Meatless Mondays
Designating one day a week as a no-meat day is a great way to give yourself a slow introduction to a plant-rich diet, and cutting back in a small way like this, can be very impactful! Tofu, jackfruit, veggie burgers, and mushrooms are great substitutes if you're looking for a protein alternative.

HELPFUL IDEAS

- **Keep a supply of <u>frozen</u> veggies in stock**
  You won't have to remind yourself to buy them constantly at the store, and you'll have them on hand without rushing to cook before an expiration date.

- **Swap out an animal-based ingredient for a plant-based**
  That could mean opting for oat milk in your morning coffee instead of creamer or using soy milk in your cereal rather than dairy. Or, try swapping butter for coconut, vegetable, or olive oil in your cooking.

- **Try out a new plant-based alternative each week**
  There are tons of varieties of veggie burgers, vegan cheeses, and even desserts: coconut milk ice cream, oat milk ice cream, etc., are fun ways to dip your toe in the plant-based world!

Find hundreds of delicious meatless recipes on line and start your **MEATLESS MONDAYS**.

CHAPTER V
REDUCE TRASH

Trash, a major component of Global Warming. The waste humans generate   has been detrimental to our environment for years. Humans are generating too much trash and cannot deal with it in a sustainable way. Waste that is not biodegradable and cannot be properly recycled is filling our oceans and landfills.

A recent study found that of the 6.3 billion metric tons of plastic waste that has been produced, only 9% of that plastic waste had been recycled

In 2017, the Environmental Protection Agency calculated that the total generation of municipal solid waste in the United States that year was 267.8 million tons. The amount of waste generated affects the environment in multiple ways: its contribution to the worsening climate change crisis, its negative impact on wildlife and the natural environment, and its detriment to our very own health and wellness.

The burning of large, open piles of trash in various parts of the world emits dangerous levels of carbon dioxide, a greenhouse gas that is a major cause of global warming. Researchers have calculated that approximately 40% of the world's trash is burned in this fashion posing large-scale risks to both our atmosphere and the people that live near these burning sites.

Ecosystems vary widely from location to location. However, one of the most important consequences of our global waste problem concentrates itself to our marine life and waterways. The people who depend on the ocean for their livelihoods, are deeply at risk of losing their cultural habitat. They cannot distinguish between fish that is good or bad.

They consume the fish, which results in their death because the aquatic animal cannot process plastic. Plastic consumption affects fish, seals, turtles, whales, and many other aquatic animals, as scientists have also found many plastic fragments in over a thousand species.

Due to ingestion of trash or plastics, starvation is usually the next step because some species do not have high acidic levels in their stomach to break down the object that they ingested. There are some animals that do but plastic fragments have been known to be able to last 100 years. Based on the overall problem with dumping in our oceans and waterways our world's species are in danger of extinction.

Human health is also at high risk with the production of TRASH. We keep producing large amounts of trash, we do not dispose of it correctly, and it ends up in the landfill that further compromises our health and wellness. The more emissions that we produce due to how much trash we generate, affects us long term. Diseases as asthma, birth defects, cancer, cardiovascular disease, childhood cancer to name a few is a rising concern based on the amount of accumulated trash we produce.

Trash attracts bacteria, vermin and insects.
As waste decomposes, it creates microbes that can cause diseases and other health issues like gastroenteritis, malaria, typhoid, cholera, as well as stomach pains, vomiting, and chronic diarrhea. Vermin and insects are attracted to the smelly trash and spread the debris and infectious diseases.

Overflowing waste bins are an ideal breeding ground for bacteria, insects and vermin. The flies that visit the garbage are also the same flies that roam around your lunch buffet and drop their off springs on your plate. By doing so, they increase the risk of you contracting with salmonella, which causes typhoid fever, food poisoning, gastroenteritis, and other major illnesses. Besides flies, other animals that thrive from the garbage in and around the containers include rats, and stray dogs.

To avoid overflow of trash, make sure your community has enough recycle trash containers.

TRASH a number one contributor to Global Warming.
Trash ends up in already overfilled Landfills. It's important to
recognize trash and how, when and where to discard it.

CHAPTER VI
REDUCE LITTER

Litter is a major contributor to Global Warming. What is Littering? Littering is the improper disposal of waste products whether it happens intentionally or unintentionally. Manufacturers began producing a higher volume of litter-generating products and packaging in the 50's made of materials like plastic. Increases in littering and illegal dumping contributed to air pollution, land pollution and ocean pollution.

Throwing a cup out a window or dropping food packaging on the ground has monumental effects on the environment. The following littering facts can educate on what littering is, why people litter, how litter affects the environment and simple solutions to make major changes to combat climate change issues.

Litter can take a variety of forms, but some items are littered more frequently than others. Researchers found that frequently littered items include:

- Cigarette butts
- Fast food packaging (bags, cups, serving items)
- Food packaging (wrappers, boxes, film, Styrofoam)
- Alcoholic beverage containers
- Tire and vehicle debris
- Plastic bags
- Plastic bottles
- Construction waste

Cigarettes are the most frequently littered item, with an estimated 9.7 billion cigarette butts discarded everywhere, scattered along green spaces, sidewalks, roadsides, beaches, waterways. Cigarettes are the most littered item on earth. worldwide, about 4.5 trillion cigarettes are littered each year

Cigarettes make up more than one-third nearly 38 percent of all collected litter. Disposing of cigarettes on the ground or out of a car is a very common activity. Approximately, $16 million is spent on cigarette clean-up.

Cigarette butts release toxic chemicals, such as arsenic (used to kill rats) and lead, to name a few, into the environment and can contaminate water. The toxic exposure can poison fish, as well as animals who eat cigarette butts. The cotton (in the filter is not cotton) 98 percent of cigarette filters are made of plastic fibers (cellulose acetate) that are tightly packed together, which leads to an estimated 1.69 billion pounds of cigarette butts winding up as toxic trash each year.
The plastic fibers in cigarettes are non-biodegradable, meaning they won't organically break down.
Estimates on the time it takes cigarettes to break down vary, but a recent study found that a cigarette butt was only about 38 percent decomposed after two years.

Plastic Gloves a littered product.
A study found that, since the COVID-19 pandemic, there has been an estimated 207 million gloves and masks littered across roadways and waterways.

The top illegally dumped items include household garbage, hazardous waste, appliances, mattresses, furniture and e-waste.
In a previous study conducted in 2009, 76% of roadway litter comes from motorists and pedestrians intentionally discarding items. Considered reasons for casual littering:

- Laziness or carelessness
- Lack of access to trash receptacles

- Lenient law enforcement
- Following the leader of litter already present in the area

People who litter out of laziness or carelessness often believe that someone else (a maintenance worker) will pick up and dispose of the trash. Others may not have been educated on the impact of littering on the environment, while others may live in an area where littering is an accepted part of the culture.

Litter has important serious consequences to the community and to the environment, other than being unsightly.

1. Litter Causes Pollution

As litter degrades, chemicals and microparticles are released. These chemicals aren't natural to the environment and can, therefore, cause a number of problems. For example, cigarette butts can contain chemicals such as arsenic and formaldehyde. These poisons can make their way into the soil and freshwater sources, which negatively impacts both humans and animals. In fact, 60% of water pollution is attributed to litter.

In addition to water and land pollution, litter can also pollute the air. Researchers estimate that more than 40% of the world's litter is burned in the open air, which can release toxic emissions. These emissions can cause numerous health and respiratory issues.

2. Litter can disrupt and kill wildlife

Every day animals are victims affected by litter. Researchers estimate that over one million animals die each year after ingesting, or becoming entrapped in, improperly disposed trash and litter.

Plastic litter is the most common killer of animals and marine animals. Each year over 100,000 dolphins, fish, whales, turtles and more drown after becoming entangled in or digesting plastic litter.

3. Litter can cause disease

Improperly discarded trash is a breeding ground for bacteria and diseases. Litter can spread diseases, viruses and parasites through two methods, direct and indirect contact.

Germs can be transmitted directly by physically coming into contact with litter. This can happen by picking up, touching, or by accidentally injuring themselves on improperly disposed of trash.

Bacteria and parasites can also be transmitted to humans indirectly through an affected vector. Vectors are animals or insects that come in contact with contaminated litter and then transmit those contaminates to humans.

4. Dispose of litter is costly

Litter along the highway must be picked up and disposed of *daily* in order to keep drivers safe. This costs taxpayers money and diverts funds from other projects like road repair. The US spends 11.5 billion annually in litter costs. Surveys have shown that homeowners, realtors and business owners believe that litter significantly decreases the value of their property.

Great simple solutions to resolve the Litter problems today. It's important to be educated on what littering is and how it affects our community and our environment.

1. **Become and Environmental Steward**
   - **Attend and set up organized cleanups.** Working to clean up your community is not only beneficial to the environment but will make your community beautiful. **Clean one littered area.** People tend to litter more in an area that already has large amounts of litter, so cleaning up litter in one specific area can have a positive impact.

- **Adopt a highway**: If you own or operate a business, consider adopting a highway, which are more regularly cleaned.
- **Make sure items in your garbage bins and vehicles are secured**: Securing your garbage bins and cargo can help you avoid unintentional littering. Check the lids of your garbage bin regularly to secure they don't fly off.

2. Decrease litter by increasing Public Disposal Bins

People litter due to a lack of public garbage bins, or overflowing bins that do not get emptied regularly. By increasing the number of available trash receptacles and the frequency they get cleaned, communities can see a reduction in littering. To increase the number of garbage bins in the community:

- **Reach out to local businesses** or put an extra bin outside your own business if you have one.
- **Call the leasing office of your apartment complex** if you notice there is not an adequate number of trash receptacles.
- **Speak up when your community is organizing an event** to ensure that enough waste collection services will be provided.
- **Contact school districts and local governments** and encourage them to invest in their community's health by increasing the number of trash bins.

3. Impose Stricter Laws and Regulations

An effective barrier to littering are strong anti-litter laws and regulations. Both individuals and businesses are more likely to follow litter laws when there are serious legal or financial consequences.

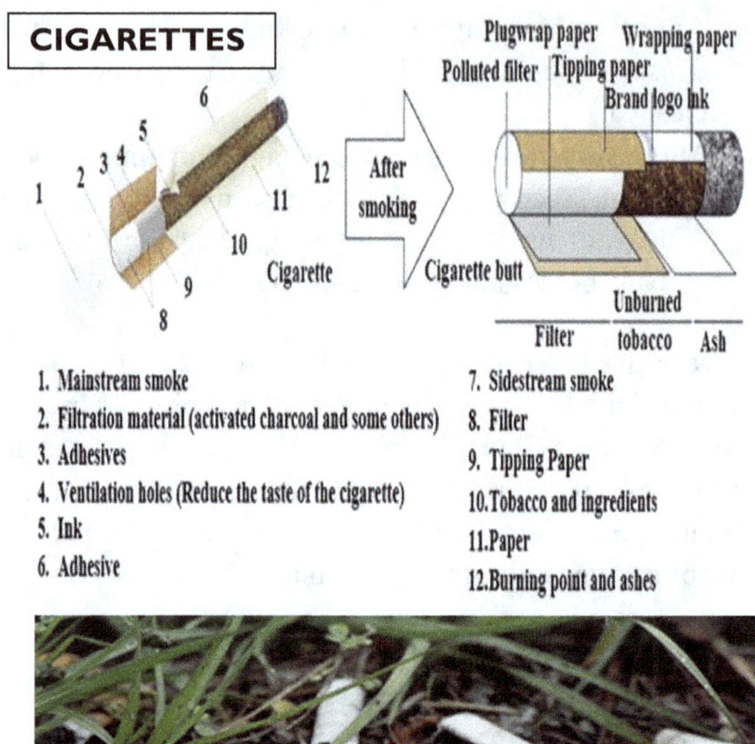

1. Mainstream smoke
2. Filtration material (activated charcoal and some others)
3. Adhesives
4. Ventilation holes (Reduce the taste of the cigarette)
5. Ink
6. Adhesive
7. Sidestream smoke
8. Filter
9. Tipping Paper
10. Tobacco and ingredients
11. Paper
12. Burning point and ashes

CIGARETTE BUTTS are the most littered item on Earth, with an Estimated 4.5 trillion discarded each year. In the United States, 9.7 billion cigarette butts are littered each year, with 4 billion of those in waterways. Cigarette butts are made of a non-biodegradable plastic called cellulose acetate. They can take decades to degrade and break down into microplastics. Cigarette butts can leach toxic chemicals into water, where they can remain for as long as 10 years. They can also pollute soil, beaches, and waterways. Studies have shown that cigarette and e-cigarette waste is harmful to wildlife.

36

CHAPTER VII
REDUCE WATER WASTE

Water is not an unlimited resource.

Conserving water goes beyond just saving water; it plays a vital role in conserving energy and reducing greenhouse gas emissions.

A FEW IMPORTANT FACTS ABOUT WATER

Ninety-seven percent of all water on the earth is salt water, which is not suitable for drinking. 3% of water on Earth is fresh water, and only 0.5% is available for drinking. The other 2.5% of fresh water is locked in ice caps, glaciers, the atmosphere, soil, or under the earth's surface, or is too polluted for consumption.

With growing population rates and such a small percentage of all the water on Earth fit for consumption, it only makes sense that we must preserve and conserve this precious resource. Water conservation means using our limited water supply wisely. It is a responsibility for everyone to learn more about water conservation and how we can help keep our sources pure and safe for future generations. There is not an endless supply of water.

CONSERVE WATER IMPORTANT FACTS

- Conserving water minimizes the effects of drought and **water shortages.** By reducing the amount of water we use, we can better protect against future droughts.

- Failing to conserve water can eventually lead to a lack of an adequate water supply, which can have drastic consequences that can include rising costs, reduced food supplies, and health hazards.

- Reducing our water usages reduces the energy required to process and deliver it to homes, businesses, farms, and communities, which, in turn, helps to reduce pollution and conserve fuel resources.

- Freshwater resources are also used for beautifying our surroundings—watering lawns, trees, flowers, and vegetable gardens, as well as washing cars and filling public fountains at parks. Failing to conserve water now can mean losing out on such uses in the future.

- Hospitals, firefighters, gas stations, street cleaners, health clubs, gyms, and restaurants all require large amounts of water to provide services to the community. Reducing our usage of water now can help continue these services.

- Reduce your carbon footprint by conserving water that saves energy. Energy is needed to filter, heat and pump water to your home.

- Conserving water will keep our wetland habitats and ecosystems healthy supporting animals like otters, water voles, herons and fish.

ADDITIONAL WATER FACTS

This chapter will examine the reasons and benefits of saving water and simple solutions to help you save more water.

Freshwater is becoming more of a scarce resource with each passing year. Population increase has caused a rise in the water demand. We need to save water to keep our daily activities going, to grow the food that we eat and to preserve natural habitats. Saving water also reduces the burden on water treatment facilities.

More importantly It also makes it possible for areas with little access to water to get enough to support their everyday needs.

Our daily requirements for water is a permanent structure to support our everyday needs.

We take water for granted, but water is a limited resource. By conserving water, more people in the world will have enough water to perform essential daily activities. Saving water ensures that we make a sufficient supply to have it available for future generations.

We use water in almost everything; it is part of our daily routines and is indispensable.

Water is needed to stay clean, take baths, and ensure a hygienic environment. The average person uses around 140 litres of water a day.

Cleaning the basics like clothes, cars, dishes, utensils all require water. We also need to drink at least a half a gallon of water daily. The body can survive without food for up to two months, but the organs begin to shut down after three days without water. To continue to perform these daily activities and keep our bodies healthy, we must look at ways to save water and avoid waste.

Plants need water too. Plants provide food for humans. When there is a water shortage, plants can't grow because they need water to transport minerals and nutrients from the soil for cooling and photosynthesis.

Agriculture is the primary source of food production, and globally, about _70% of the world's fresh water is used to irrigate crops_. When there is a drought, plants will die, and it becomes difficult to grow food.

Cultivating the habit of saving water means having sufficient water to grow food and sustain life even during droughts. It also ensures that plants can survive and the ecosystem is protected.

Wastewater affects the environment. About 80% of the world's wastewater goes back into the environment without being treated, which means about 1.8 billion people use a contaminated source of drinking water.

Wastewater is a product of agricultural, industrial, and domestic household practices. Wastewater's acidity has harmful effects on the environment. When it flows into streams in large quantities, it can raise its temperature, leading to the disturbance of the natural balance of aquatic life.

When you use a lot of water, you increase the quantity of wastewater that goes back to the environment untreated. For example, 62% of indoor water use comes from faucets, toilets, and showers. According to the United States Environmental Protection Agency, about 23,000 to 75,000 sanitary sewers are overflowing each year.

Adopting innovative water saving techniques can significantly reduce the amount of wastewater that goes into the environment.

There is a limited fresh water supply. Only 2.5% of all water on earth is freshwater, and less than 1% of this is accessible for consumption. People must begin to think about the sources of water and how much is available for use. It is believed that we have an unlimited access to water.

Water is becoming less accessible as the world population continues to increase. Climate change will magnify these conditions.

Saving water prevents groundwater depletion. Pumping groundwater faster and more frequently before it can renew itself leads to an alarming shortage of groundwater supply.

Half of the total population relies on groundwater for drinking water. In addition, it provides over 50 billion gallons per day for agricultural needs.

Turning off the faucet, reducing the use of dishwashers and washing machines, and cutting down unnecessary use of water, are simple ways you can save water.

Conserving water reduces demands on technology and infrastructure. According to the United States Environmental Protection Agency, 34 billion gallons of wastewater are treated by wastewater facilities in the United States daily. In the same vein, approximately 20 percent of homes in the United States use septic systems that locally treat their wastewater.

Saving water saves you money. An average American household uses over 300 gallons of water per day at home. Nationally, outdoor water use accounts for 30% of household use.

You can adopt simple water-saving tips to save money and water. For example, you can invest in an eco-friendly shower equipped with a multi-sensor perception technology that allows you to control your water usage while enjoying a pleasing shower experience.

Conserving water saves energy. The pumping of water requires high amounts of energy. Water travels hundreds of miles through pipelines before it get to you. You can reduce your carbon footprint and save energy by simply using less water.

Wildlife and aquatic life needs water to survive. Water is home to countless species. It also serves as protection from natural hazards and a source of drinking water. Toads, salamanders, and frogs live in water. Butterflies extract valuable minerals from muddy water. Riverdolphins and diving bell spiders rely on freshwater to survive.

SAVE WATER IN THE KITCHEN
- o   Put a large bottle of tap water in the fridge to save waiting for the tap to run cold. Waiting for the tap to run cold can waste 10 litres of water a day!
- o   Only fill the kettle with the amount of water needed.
- o   Put lids on saucepans to reduce the amount of water lost during heating.
- o   Put your dishwasher and washing machine on with full loads and on an eco-setting wherever possible.

SAVE WATER IN THE BATHROOM
- o   Turn the tap off while brushing your teeth. A running tap can waste more than 6 litres of water a minute!
- o   Purchase a water-efficient toilet
- o   Shower instead of bathe. An average bath uses around 80 litres of water, but a shower typically uses between 6 and 45 litres.
- o   Install water-efficient taps and showers to minimize heating water this will save you money on your water and energy bills, as well as decreasing your carbon footprint.
- o   Fix a dripping tap. A dripping tap can waste 15 litres of water a day!

SAVE WATER IN THE GARDEN
- o   Sprinklers can use as much as 1,000 litres of water an hour!
- o   Use a water barrel to catch large amounts of rainwater and use this to water your plants, clean your car and wash your windows.
- o   Use mulch and bark in your garden, it will help to reduce evaporation by up to 75%.

SAVE WATER IN THE HOME

- o **Stop leaks.** Check all water-using appliances, equipment, and other devices for leaks. Running toilets, steady faucet drips, home water treatment units, and outdoor sprinkler systems are common sources of leaks.

- o **Replace old toilets**. The major water use inside the home is toilet flushing. If your home was built before 1992 and you haven't replaced your toilets recently, you could benefit from installing a Water Sense labeled model that uses 1.28 gallons or less per flush. A family of four can save 16,000 gallons of water per year by making this change.

- o **Replace old clothes washers**. Washers are the second largest water user in your home. If your clothes washer is old, you should consider replacing it with an ENERGY STAR certified clothes washer. Most ENERGY STAR clothes washers use four times less energy than those manufactured before 1999. To save more water, look for a clothes washer with a low water factor. The lower the water factor, the less water the machine uses. Water factor is listed on the certified product list.

- o **Plant the right plants.** Whether you're installing a new landscape or changing the existing one, select plants that are appropriate for your climate. Consider landscaping techniques designed to create a visually attractive landscape by using low-water and drought-resistant grass, plants, shrubs, and trees. If maintained properly, climate-appropriate landscaping can use less than one-half the water of a traditional landscape.

- o Provide only the water plants need. Automatic landscape irrigation systems are a home's biggest water user. To make sure you're not over watering, adjust your irrigation controller at least once a month to account for changes in the weather. Install a rain shutoff device, soil moisture sensor, or humidity sensor to further control irrigation.

Food can't grow without water

Humans and animals need food. Plants provide food for both. When there is a water shortage, plants can't grow because they need water to transport minerals and nutrients from the soil for cooling and photosynthesis. As the world population continues to increase, water is becoming less accessible. Moreover, with bio-energy demands and climate change, this water crisis is likely to be amplified.

By 2050, researchers estimate that nearly half of the world's population will live in water scarcity areas. By the end of the century, water deficit is projected to affect about 700 million people globally. These statistics point to one thing. We need to conserve water. Saving water prevents groundwater depletion Pumping groundwater faster and more frequently before it can renew itself leads to an alarming shortage of groundwater supply.

In the United States, groundwater is a valuable resource. Half of the total population relies on groundwater for drinking water. In addition, it provides over 50 billion gallons per day for agricultural needs. Unfortunately, today, many areas across the United States are experiencing groundwater depletion.

We also run a risk of the water quality deteriorating. The more the grounds are dug for water, groundwater can be contaminated by saltwater intrusion. This is because continuous water pumping can disrupt the boundary between freshwater and saltwater, leading to saltwater migrating inland. When groundwater is contaminated by saline water, it becomes unusable.

Conserving water reduces demands on technology and infrastructure.

According to the United States Environmental Protection Agency, 34 billion gallons of wastewater are treated by wastewater facilities in the United States daily. In the same vein, approximately 20 percent of homes in the United States use septic systems that locally treat their wastewater.

The more water is used, the more pressure is placed on wastewater treatment facilities. For example, when septic tanks are overburdened by water usage, it can result in water leaking. This untreated water from the leak will have adverse effects on the surrounding soil. When these wastewater treatment facilities are overloaded with too much water in a short period, their lifespan can be shortened, leading to system failure. The more water is used, the more these facilities wear down and require replacement. Saving water reduces the demands to create and maintain water treatment and delivery facilities such as septic tanks and sewage plants.

Rain barrels can hold anywhere from 15 to 5,000 gallons of water. The most common sizes for residential use are 55 gallons.

REDUCE EMISSIONS. A MAJOR CONTRIBUTOR TO GLOBAL WARMING.

Vehicle emissions contribute to the formation of ground level ozone (smog), which can trigger health problems such as aggravated asthma, reduced lung capacity, and increased susceptibility to respiratory illnesses, including pneumonia and bronchitis.

Air pollutant emissions emitted by motor vehicle exhausts include:

carbon monoxide (CO),
nitrogen oxides (NOx),
rticulate matter (PM); and.
volatile organic compounds (VOC).

CHAPTER VIII
REDUCE EMISSIONS

What are emissions? Where do they come from?

Emissions is the term used to describe the gases and particles which are put into the air or emitted by various sources.

The amounts and types of emissions change every year. These changes are caused by changes in the nation's economy, industrial activity, technology improvements, traffic, and by many other factors. Air pollution regulations and emission controls also have an effect. The National Air Pollutant Emission Trends report summarizes long-term trends in emissions of air pollutants and gives in-depth analysis of emissions for the current year. The report also discusses emission evaluation and predictions.

Criteria Pollutants

The United States Environmental Protection Agency (EPA) is mainly concerned with emissions which are or could be harmful to people. EPA calls this set of principal air pollutants, criteria pollutants. The criteria pollutants are carbon monoxide (CO), lead (Pb), nitrogen dioxide(NO_2), ozone (O_3), particulate matter (PM), and sulfur dioxide (SO_2). There are also a large number of compounds which have been determined to be hazardous which are called air toxics.

There are many sources of emissions. These have been grouped into four categories: point, mobile, biogenic, and area.

- Point sources include things like factories and electric power plants.
- Mobile sources include cars and trucks, lawn mowers, airplanes and anything else that moves and puts pollution into the air.

In 1970 The United States Congress passed Clean Air Act(CAA) which set into motion a nationwide effort to improve the country's air quality. Measuring, reporting, and using emissions data.

In order to make improvements in the air quality, the amount of pollutants in the air must be measured. The Emissions Measurement Center develops standards and evaluates testing methods so that regulations can be developed and enforced.

What is an Emission Factor?

An emission factor is a relationship between the amount of emissions that are released and the activity of the producer. Emission factors are used to predict emission levels for different industries.

What are Emission Inventories?

Emission inventories are quantities of pollutants measured over time. Emission inventories can be compared with air pollutant levels in an area to determine if increased emissions decreases the air quality.

Once the measurements are made the information must be collected and stored so that it can be used to evaluate the air quality and effects of the regulations.

Net zero emissions can be achieved by removing as much greenhouse gas from the atmosphere as what's emitted, so the net amount added is zero.

To do this, countries and companies will need to rely on natural methods – like planting trees or restoring grasslands – to soak up carbon dioxide, the most abundant greenhouse gas we emit, or use technology to capture the gas as its emitted so it doesn't enter the atmosphere.

Negative emissions is the situation where the amount of greenhouse gas removed from the atmosphere is actually *more* than the amount humans emit.

Carbon sinks
This is a reservoir that absorbs carbon dioxide from the atmosphere and locks it away. Natural sinks like trees and other vegetation remove CO2 from the atmosphere through photosynthesis.–Plants use the carbon dioxide in the atmosphere to grow. The ocean is also a major carbon sink because of phytoplankton which, as a plant, also absorbs carbon dioxide.

Scientists say preservation and expansion of natural sinks – like the Amazon Rainforest – arc crucial to reducing emissions. There are also artificial carbon sinks that can store carbon and keep it out of the atmosphere

Carbon capture and storage
Technology to remove and contain carbon dioxide from the atmosphere is known as carbon capture and storage. Carbon is usually captured at source – directly from coal, oil or gas as it burns – but new technology is being developed to literally suck carbon from the ambient air.

In both cases, the carbon can be stored, usually buried in reservoirs underground or below the floor of the sea, in what are known as artificial carbon sinks. Some scientists warn that it could be risky to inject so much carbon underground, and this process isn't currently used on a large scale. Carbon dioxide capture and storage is a process in which CO2 produced by heavy industry or power plants is collected directly at the point of emission, compressed and transported for storage in deep geological formations.

- Carbon capture, utilization and storage (CCUS) refers to the collection of CO2 from industrial sources, which is then used to create products or services, such as manufacturing fertilizer or in the food and beverage industry.
- Direct air capture and storage (DACS, DAC or DACC) is a chemical process which removes CO2 directly from the air for storage. There were 15 direct air capture plants operating worldwide. Solutions that can reduce greenhouse gas emissions, or remove them from the atmosphere, to ease the consequences of climate change.

Using fossil fuels like coal, oil and gas more efficiently for industrial processes, switching from coal and gas to renewable energy sources such as wind or solar power for electricity, choosing public transport to commute over private vehicles that run on gasoline, and expanding forests and other means of absorbing carbon.

The availability of electricity generated by wind and solar, people will be expected to start buying electric vehicles in greater numbers. Plug-in hybrid electric vehicles, which are mostly powered by a battery charged from an electrical source have an internal combustion engine to allow longer travel time.

Biodiversity
Biodiversity refers to all the Earth's living systems, on land and in the sea.

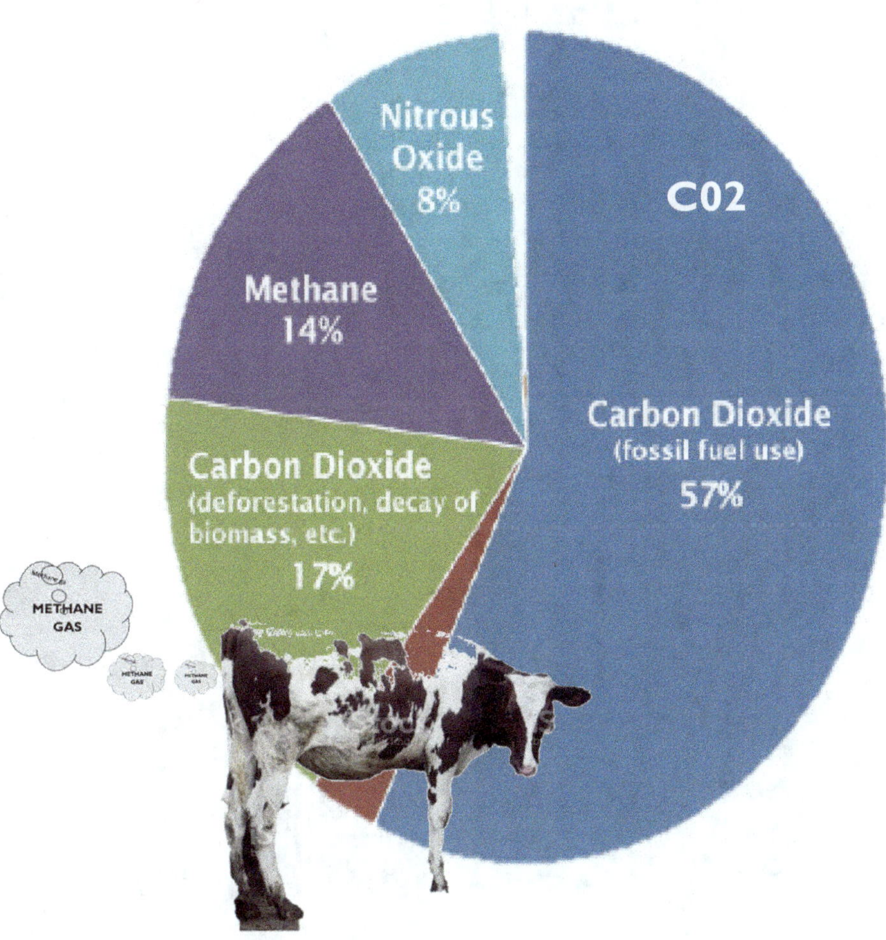

The diagram shows Carbon dioxide from fossil fuel, Carbon dioxide from deforestation and Methane gas from farm animals.

PART

B

"REUSE"

REUSE YOUR CEREAL BOX. MAKE THEM INTO BOOK AND PAPER ORGANIZERS. HOW TO MAKE YOUR ORGANIZER ON PAGE 78.

PART B
"REUSE"

CHAPTER I

REUSE

Reuse is the act of using something again, either for its original purpose or for a different function. For example, you can reuse a plastic bag from the grocery store many times. You can also reuse old clothing as cleaning rags. Reuse is different from recycling, which is the process of breaking down used items to make raw materials for new products.

Examples of reuse:
Using jars from grocery store foods to store leftovers or to take lunch to work
Using empty plastic bottles in a science experiment

Reuse is the practice of finding new ways to use things that would otherwise be thrown out. Reuse can help save time, money, energy, and resources and may also help in the following ways.

- Conserve resources

- Reduce the waste stream

- Cause less pollution than recycling or making new products from virgin materials

- Make needed items available to those who can't afford to buy them new

- Prevent solid waste from entering the landfill

- Increase the material, educational, and occupational wellbeing of citizens

- Reuse can also generate jobs and business activity that contribute to the economy.

Reuse is one of the three R's of waste management, which are: Reduce, Reuse, Recycle.

Waste management and minimization can be achieved by focusing primarily on the first of the 3 R's, "reduce," followed by "reuse," and then "recycle".

A FEW "REUSE" IDEAS

MAKE YOUR OWN WATERING CAN
Take a refill bottle. Give it a good rinse, punch some holes in the cap and one at the top of the handle, then fill it up with water and watch your garden grow. You can do this with any bottle that has a handle a pint-sized milk bottle is great for little gardeners.

USEFUL EGG CARTONS
Empty egg cartons can be colored with a splash of paint, then used in your drawer as an organizer for pens, picture hooks, safety pins paper clips etc.

A HOMEMADE FACIAL SCRUB
When you're done with your morning coffee, blend your coffee grinds with melted coconut oil and some sugar to create a naturally exfoliating face and body scrub.

REUSE OLD BEDDING
Worn-out bedding, blankets and towels can go through the cutting machine (scissors) and make endless bags a rags to use for cleaning around the house. Any fabrics that are 100% natural can even be cut up and composted.

GROW AN AVOCADO TREE
Simply wash and dry your avocado pear seed, fill a jar with water (leaving a little room at the top), then pierce the seed with four toothpicks and place it in the rim of the jar so that the broad end is

covered by around an inch of water. Place your jar in a warm, sunny spot, out of direct sunlight and replenish the water as needed. In two to six weeks you should see sprouts and a stem emerge. Once the stem gets to six inches, trim it by half. Then, once new leaves come back and the roots are sturdy, you can plant your tree in a pot.

UPCYCLE OLD GLASS BOTTLES INTO REED DIFFUSERS

Empty glass bottles, old beer bottles or perfume bottles will work perfectly. Just blend your favorite essential oils with a base like sweet almond oil, then pop in a handful of diffuser sticks or bamboo skewers and let the beautiful fragrances linger.

MAKE YOUR OWN PLANT POT

Plants need repotting? Craft your own colorful plant pot. Draw a ring around the bottle just above the label works well - then simply cut along the line and for drainage, use your scissors to punch a few holes in the base.

USE MAGAZINE PAGES FOR WRAPPING PAPER

Put a unique (and sustainable) twist on thoughtful gifts by swapping wrapping paper for old magazine pages. This one doesn't need much instruction, but with the simple addition of some colorful ribbons or hand drawn designs, your gift wrap will be greatly appreciated and a cherished presentation.

MAKE A TOTE BAG FROM OLD JEANS

Declutter your wardrobe. Get your old denim, and your tools for cutting and sewing, and make your TOTE BAG.

REFILL EMPTY JARS

Use your very useful glass jars to store your dry goods, like rice, pasta and lentils.

MAKE REUSABLE BAGS

Plastic bags have a large environmental footprint.

Prevent Pollution.

Save Money.

Better Quality

Suit Your Style

Declutter Your Space

Benefits of using Reusable Grocery Bags?

Reusable bags can ease the plastic bag pollution and the effects of plastic has on the environment.

From their production to their lack of recyclability to their tendency to end up in landfills and even worse, the many years they take to decompose. Plastic bags pose a major threat to the health and wellbeing to humans and our planet.

Paper bags avoid some of the pitfalls of plastic bags because paper bags can be recycled and take less time to decompose if they don't make it to a landfill. Even so, paper bags are made either from trees, which are important to conserve, or from recycled material, which takes a lot of energy to produce.

WHY REUSABLE BAGS

1. Conserve Resources

Plastic bags have a very larger environmental footprint. It takes an enormous amount of energy to make them. Twelve million barrels of oil are used to manufacture the plastic bags consumed in the United States each year,

2. Decrease Pollution

The effects of plastic on the environment is devastating. Plastic products such as plastic bags can take between 15 and 1,000 years to break down, and that's assuming they even make it into a landfill instead of winding up in water such as streams, rivers or the ocean. Of the 100 billion plastic shopping bags Americans use each year, only about 1 percent are recycled, so a lot of plastic bag pollution is generated annually.

3. Avoid Recycling Problems

Other problems with plastic bags. Bags get snagged on conveyors belts and wheels, clogging the machinery; they can be difficult to separate from other recycled products; or they end up drifting to other parts of the recycling plant or end up outside of the plant.

4. Protect Wildlife

More than 100,000 marine animal deaths are caused each year when marine animals mistake plastic shopping bags in the ocean for food. Plastic bags also get snagged in trees, and small animals can become trapped in them, leaving to even more wildlife deaths in the environment.

5. Enjoy Strength and Durability

Unlike plastic bags, reusable bags are durable. Reusable bags are easier to use for both loading and unloading groceries, and your purchases are more likely to survive the trip to and from the store. Leaks are less likely to be a problem with reusable bags, and it's easier to control where reusable bags have been and what germs they may carry.

6. Save Money

Many stores in the United States now charge customers for plastic bags. While a nickel or dime per bag may not feel like a lot to spend, consider how quickly that adds up for a consumer who uses five or ten bags with each trip to the store. Not to mention, the average American family uses nearly 1,500 plastic shopping bags each year, according to grocery shopping statistics.

7. Repurpose For Other Uses

Reusable grocery bags can be used for a lot more than carrying groceries. A nice reusable bag is essentially a structured tote bag, and can be used to carry your lunch, pack goods for outings, and more.

8. Reusable grocery bags have value

Reusable grocery bags are far more environmentally-friendly due to the fact that they don't end up in landfills or garbage units nearly as often as regular shopping bags.

Reusable grocery bags can be used for carrying far more things that regular grocery bags. As long as you wash them properly and store them in a clean place, away from excess sunlight or humidity, your reusable grocery bag can last you for years.

How do you disinfect reusable grocery bags?

If you don't have time to wash your reusable grocery bag thoroughly, then you can simply disinfect it. An antibacterial wipe works just as well.

How long do reusable bags last?

You can expect a high-quality reusable tote bag to last for at least one year A tote bag may also last significantly longer if used cautiously, for lighter weighing goods.

Can reusable grocery bags go in the dryer?
It is not recommend to put reusable bags in the fryer. Putting them in the dryer will only tear their fabrics much faster than they would tear from the weight of carried groceries.

SUGGESTED REUSABLE PRODUCTS

REUSABLE BAMBOO PAPER TOWELS

REUSABLE PAPER TOWELS

One of the most feasible solutions to our world's paper towel problem is using reusable paper towels. Unlike traditional paper towels, reusable towels give more usage per towel and require less resources to be produced, resulting in less deforestation, less pollution, and less water and energy usage.

HOW REUSABLE PAPER TOWELS WORK

Reusable paper towels are similar in concept to regular dish towels. Made from cotton, cellulose, or another durable material, these reusable towels are made more durable than regular paper towels and are meant to be used for multiple kitchen tasks before being washed in a dishwasher or by hand and then being left air dry like any other kitchen cloth.

BAMBOO PAPER TOWELS

Bamboo is a sustainable paper source that many paper products have switched to. Cellulose dishcloths, Microfiber towels.

Bamboo paper towels are the better alternative to paper towels. They are eco-friendly and do not cause any harm to the environment. It saves the Earth from depleted ecosystems and **deforestation.**

Super Bamboo Paper Towel (1 Roll) 100% Bamboo Fiber - Absorbent, Durable and Reusable up to 80 Times, 1 Roll Can Lasts up to 6 Months. The unique growing properties of bamboo, along with the ethical and eco-friendly manufacturing process, means bamboo toilet paper is environmentally-friendly and has a lower carbon footprint. Bamboo provides an incredibly durable yet biodegradable material for use.

Bamboo toilet paper is a biodegradable and environmentally friendly product. It's often made from a combination of bamboo pulp and sugar cane. It doesn't contain any chemicals or dyes that could destroy the bacteria in septic systems.

RE-USE YOGURT CUPS TO START YOUR WINDOW SILL GARDEN.

Steps on how to make a reusable bag from an old Tshirt on page 72.

PART C

"RECYCLE"

PRODUCTS MADE FROM RECYCLED PLASTICS
1. Shampoo Bottles 2. Film And Sheeting 3. Traffic Cones
4. Packing Materials 5. Trash Bags 6. Kitchenware
7. Countertops 8. Carpeting

PART C
"RECYCLE"

CHAPTER I
RECYCLE

To **recycle** is to submit items to be broken down into base materials and converted into future products.

Here are a few ways recycling helps the environment and fight climate change:

RECYCLING CONSERVES RESOURCES

1. Recycling Conserves Resources

When we recycle plastic, we reduce the need for more plastic to be manufactured. By recycling paper, we do our part to lessen deforestation and **72.** Separating cans and other metals helps to cut down on mining and our growing need for raw materials.

2. Recycling Saves Energy

It takes much more energy to create industrial-grade materials from scratch than it does just to reform old materials and reuse them. For example, it is **estimated** that "recycling aluminum saves 90% to 95% of the energy needed to make aluminum from bauxite ore." Recycling saves energy because recycled materials don't require factories to expend so much greenhouse gas emissions than they would if they had created the same item again from scratch using raw materials.

3. Recycling Protects the Environment

When we cut down on the amount of new materials and natural resources we need to extract from the earth, whether through

63

farming, mining, logging, etc., we protect vulnerable ecosystems and wildlife from harm or eradication and allow them to exist for generations to come.

4. Recycling Slows the Spread of Landfills

In the United States alone, there are **2,000 active landfills** all full of solid waste and emitting greenhouse gases into the atmosphere as it all slowly decomposes. This doesn't account for the many that have been closed due to reaching capacity, or the future landfills that'll have to be created when the active landfills become land*full*.

5. Recycling Creates Jobs

In the US, the Natural Resources Defense Council (NRDC) **found** that reaching a 75% recycling rate nationwide would create 1.5 million new jobs.

6. Recycling Supports the SDGs

Our future generations will depend on established sustainable goals.

Here are the most common materials to recycle and helpful steps to accomplish the goal.

1. Plastics

Plastics **are the absolute worst**. The plastic grocery bag takes 10–20 years to decompose. That single-use water bottle takes up to 450 years to break down in a landfill. Other, more durable plastics? Up to 1,000 years! Plastics are challenging for the environment. Plastics are difficult to recycle. Not all plastics can be recycled. There are different types of plastics which require varying processes and considerations. The Resin Identification Coding System (**<u>RIC</u>**) separates plastics into seven different types.

SEVEN TYPES OF PLASTICS:

1. Polyethylene terephthalate (PETE or PET)

2. High-density polyethylene (HDPE or PE-HD)

3. Polyvinyl chloride (PVC or just V)

4. Low-density polyethylene & Linear low-density polyethylene (LDPE & PE-LD)

5. Polypropylene (PP)

6. Polystyrene (PS)

7. Other plastics

The number above also corresponds to the recycling number of a plastic, which you can most often find on the bottom of the item, along with an abbreviation (in parentheses above).

When you recycle plastics, you cut down on the long degradation time, while helping to make sustainable new products. According to Researchers "recycled plastic bottles are now being used to make carpets, clothing and even auto parts."

2. Paper

Paper products are relatively simple to recycle. Depending on your country, city, and district regulations, you may separate all paper together, or separate simple paper products from plastic-coated paper products, such as a coffee cups or orange juice carton. It takes special chemicals and an extra step to remove the glue, plastic, and other residue from the paper in order to reuse it.

3. Glass

Glass is just about the longest-lasting man-made material, taking up to **1 million years** to degrade in the environment! There is also the loss of sand. Sand is the key ingredient in most glass, but the supplies are dwindling.   Scientists **are saying** we're facing a growing sand shortage, some going as far to call it a sand crisis.

Glass is one of the easier materials to recycle. First, a treatment plant sorts them by colors. Then, they give them a wash and remove stickers and other impurities. Finally, they melt down crushed glass pieces and shape them into new bottles and jars ready for us to purchase again.

4. Metal

Metals must be mined from the earth, which damages the areas and environments where mining activity occurs. Most metals can be recycled together, as recycling plants sort them into their respective categories.

5. Organic Materials, Food & Compost

Organic waste such as food is the most biodegradable of the lot. The best way to recycle your own organic waste is to start composting it. Compost is organic material that has broken down, and it may appear to you as rich, dark soil.

Making compost is simple, and all you need is the passage of time. Then, when you've transformed past eggshells and orange peels into nutrient-rich dirt, use it for planting or gardening.

6. Electronics

Known as *E-waste* (short for *electronic waste*), this includes all discarded electronic items, whether broken, unwanted, or at the end of their working lives.

The hard part of recycling electronics comes down to their constituent parts. There are dozens of intricate wiring on the average circuit board, made out of a variety of metals, epoxy, glass, and other materials

Almost all of the components can be reused. **According to the EPA**, "for every million cell phones we recycle, 35 thousand pounds of copper, 772 pounds of silver, 75 pounds of gold and 33 pounds of palladium can be recovered."

To recycle your used electronics, there are often bins at large retailers, where you can drop off your e-waste,

7. Batteries

Batteries require special consideration when discarding them, as they contain toxic chemicals and heavy metals which shouldn't decompose at your average landfill. There's a potential for great harm to the environment should batteries end up there. There is also a lot of value in recycling batteries.
There are used battery receptacles located at many stores in the community.

8. Tires and Commercial Rubber

When a tire goes flat or the tread wears off its ready for recycle. Not only are tires a lot of waste, they take up lots of room at landfills.
Many tires, and other commercial rubber materials, are difficult to recycle. Much of them have been burned to get rid of them, even when properly discarded. However, as technology advances, there have also been advances in the materials used for tires, with an increasing amount of biodegradability and reuse potential.

9. Clothing & Textiles

Recycling clothing, rags, sheets, curtains, linens, and other similar materials, is a key way to reduce municipal solid waste. The U.S generated 16,890 tons of textiles in 2017, of which just 2,570 tons were recycled.

If they make their way to landfills, clothing and other fibrous materials can take up to several hundred years to break down. But, recycling these textiles helps the environment by skipping the landfill and sending the clothing to plants to be sorted, cleaned, shredded, and purposed.

10. Fiberboard & Paperboard

Paperboard is thin and formed of one layer, like paper, but thicker, less foldable, and more rigid than paper (think of a greeting card). *Corrugated fiberboard* is the three-layer kind you may be familiar with in shipping boxes, consisting of two rigid layers sandwiching a wavy middle one for strength.

In some recycling programs, there's a differentiation between fiberboard recycling and paperboard recycling—some accept paperboard with paper, others accept paperboard with fiberboard separate from paper, and a few want all three to be separated.

From reducing carbon emissions to conserving natural resources, recycling is one of the best ways we can fight climate change.

CONSIDER THESE RECYCLE PROJECTS
Donate old jeans to a local thrift store.

Use old socks to make dusting pads, rags, or toys. You can also use them to cover drafty spots in your home. You can also compost old socks if they are made of natural fabrics.

Bras are often made of **synthetic fibers**, as well as many components of varying materials, and could literally take hundreds of years to fully biodegrade. About 85 percent of textiles produced in the U.S. end up in landfills on an annual basis. Therefore, finding a proper bra disposal method is absolutely crucial.

Old pillows are reused by using their stuffings to make other items, such as carpets, mats, and insulation. You can also drop off the exterior fabric of used pillows at local charities, community drop-off bins, private clothing recyclers, local transfer stations, and special textile recycling events. You can also compost some pillows if they are made of natural materials, such as 100% cotton, wool, or down.

Recycling old shoes to your local charity shop. They can be repaired and reused. Before you drop off your shoes, check them thoroughly for holes, a missing sole, or look otherwise past their useful life.

PLASTICS, RECYCLED INTO TRAFFIC CONES

IMPORTANT GREEN UPDATES

USE GREEN PAINT

BENEFITS OF USING ECO-FRIENDLY PAINTS FOR YOUR HOME

Today's eco-friendly paints are available in many stores. This green paint has no strong scent and no harmful effects on people, plants, animals, or the air.

BENEFITS OF USING GREEN PAINT
1. Fewer VOCs

Eco-friendly paint has fewer Volatile Organic Compounds or VOCs and less "paint smell." Green paint can be identified by its lack of smell. According to the EPA, VOC levels are ten times more potent and dangerous inside your home.

One of the main benefits of eco-friendly paint is a lower number of dangerous VOCs.

2. Better Air Quality Inside and Outside

Air quality indoors and outdoors is less toxic when eco-friendly paints are used. A neighbor on your street who is painting with VOC paint emits many chemicals into the air that are toxic.

Side effects of breathing these toxins can range from headaches and nausea to kidney damage and loss of memory.

When you use eco-friendly paint indoors or outdoors, you are preventing the air from becoming more toxic from your use of GREEN paint.

3. Eco-Friendly Paint Looks Good on Walls

Eco-friendly paint looks as good as paint with higher amounts of VOC on your walls. There are many excellent water-based eco-friendly paints made by various companies. These paints have no negative environmental impact. They are made in many shades to complement any décor.

This is especially true if you have prepared your walls by removing flaking paint and making other repairs before you paint.

4. Avoid Toxins with VOC Paint

Your health may be better if you use eco-friendly paint. VOC paint can remain in the environment for years after it dries. Lung cancer is higher in people exposed regularly to VOCs.

5. Made of Natural Substances

Eco-friendly paint is made from natural substances. It has no toxic chemicals like VOC paint but is made from lemon peel and other natural ingredients. The manufacturing of green paint is also friendly to the environment with safe emissions.

6. The Additional Cost Is Worth It

Green paint is made to be safer for humans and the environment. VOC paints cost less but can cause health problems for humans and hurt the ozone. People who care about their health and the environment don't mind paying more because they see it as a long term health investment.

7. They Are Safe for Everyone

Eco-friendly paints can be used by kids or by a person with asthma. Since they are all-natural, there are no harmful fumes breathed in as you paint.

8. These Paints Are Vegan

Standard VOC paint is made with animal parts. Some paints contain crushed insects or gelatin from bones. They are then tested on animals after production to test for negative effects.

STEPS TO MAKE A REUSABLE BAG
FROM AN OLD T-SHIRT

You can make a reusable bag from an old t-shirt. Turn a t-shirt inside out and sew a zigzag stitch across the bottom of the shirt opening. To create more finished looking corners, you can add a zigzag stitch at a diagonal on each corner. Cut 2 1" strips for your handles (determine the length) Sew them in on the rim. Zig zag the closures. And you have your REUSABLE BAG. You can add what you wasn't to make it a stronger bag.

SIMPLE STEPS TO MAKE YOUR REUSABLE BAG

1. Decide on the size of your bag and how much fabric you need.

2. Cut the fabric.

3. Sew the outer fabric, lining, handles, and top.

4. Hem the top edge.

5. Add reinforcing fabric to the bottom.

6. Sew the side seams.

7. Sew the bottom shut.

TRIVIA TEST ON REDUCE, REUSE, RECYCLE, FIND THE CORRECT ANSWERS IN PART A, PART B OR PART C.

1. Which is the best alternative to a plastic water bottle?

- Plastic water bottles are the best choice
- Plastic cup
- Metal water bottle
- Paper cup

2. What should you do with your computer when you're not using it?

- Put it in sleep mode
- Leave it on
- Turn it off but leave the wall switch on
- Turn it off and unplug it

3. What's best when it comes to lights?

- Leave them all on.
- Turn them off when you leave a room.
- It depends on the time of day.
- It depends of the time of year.

4. **What is the name for the alternative to gasoline that can be made from corn?**

- ○ Corn gasoline
- ○ Methane
- ○ Ethanol
- ○ You can't make fuel from corn.

5. **Which of the following is a renewable energy source?**

- ○ Gasoline
- ○ Wind
- ○ Oil
- ○ Coal

6. **What should you do with computers that do not work anymore?**

- ○ Throw them away in the regular garbage collection bin
- ○ Pull them apart and use some of the pieces then throw the rest away
- ○ Burn them
- ○ Recycle them properly

7. If you want to be really environmentally friendly, where should you buy your produce?

- Hannaford
- Supermarket
- A farm stand in your town
- Online

8. How full should your washing machine or dish washer be before you run them?

- Whatever you have at the time
- Halfway
- Completely
- Mostly

9. What should you do with paper products like newspaper, lined paper or computer paper when you're done with them?

- Throw them away
- Burn them
- Recycle them only when it's convenient for you
- Always recycle them

10. It is more eco-friendly to print on both sides of a piece of paper.   True   False

IN SUMMARY

1. Why should you recycle?

Recycling turns materials that would otherwise become waste into valuable resources.

2. How do hybrid cars work?

A hybrid car has elements of both gasoline powered engines, and electric powered engines, which are combined in such a way that gas mileage is increased and pollution is decreased.

3. What do people recycle the most?

E-waste is the most recycled products.

4. What is energy conservation?

Energy conservation is simply using less energy.

5. Where can I recycle batteries?

Go to Earth911.com to find battery recycling centers in your area.

6. What is the cheapest way to go green?

The best way to go green is also the best way to save money -- use less!

7. What percent of people recycle?

It is estimated that only 70% of the U.S. population recycles.

8. Which city recycles the most?

San Francisco ranks the highest on many survey-based studies on recycling and sustainability.

9. How will future generations be affected if we don't recycle?
Higher fuel prices, increased consumer debt, dirty air, and decreasing wildlife population

10. How does wind power work?
The wind turns the blades of a turbine, which in turn spin a shaft that is connected to a generator. This produces kinetic energy, which the generator uses to produce electricity.

3 STATES IN THE US HAVE THE MOST WIND TURBINES. TEXAS (17,813 MW) IOWA (6,212 MW) CALIFORNIA (6,108 MW)

IS SOLAR OR WIND MORE EFFICIENT?
Wind is a more efficient power source than solar. Compared to solar panels, wind turbines release less CO_2 to the atmosphere, consume less energy, and produce more energy overall. In fact, one wind turbine can generate the same amount of electricity per kWh as about 48,704 solar panels.

SIMPLE STEPS TO MAKE PAPER ORGANIZERS OUT OF CEREAL BOXES

Gather materials:

1. Cut off the top tabs of the cereal box.

2. Trim away excess paper, leaving a 1/2" border around the perimeter.

3. Fold the paper around the box, starting at the bottom.

4. Glue the paper onto the box.

5. Fold the extra paper over each side and glue it down.

6. Let the organizer dry.

7. Decorate the box with decorative paper or duct tape.

8. Cut and stick decorative contact paper onto the box.

9. Use double-sided tape to line the box.

10. Add a strip of glue to the magnet tape and attach to the center back of the cereal box.

11. Set the organizer aside and allow it to fully dry before using.

RENEW

Renew your commitment to restore great health and wellness for yourself and planet earth. Mother earth is fighting back and sending messages that needs our utmost attention. Our planet has been abused to the state where we must make a commitment now to restore planet health for longevity. Commitment is a difficult task but we can make small changes that will make big impacts on global warming. Make an attempt to try a few of the changes you can commit to and see the power of your commitment change climage change and global warming.

Try biking often to reduce emissions produced by cars, try meatless mondays to offset the production carbon monixide produced through production of the meat industry, try recycling instead of new products to reduce emmissons for new production and try single use bottles to lessen the dumping of plastics in our oceans and our landfills try planting a tree to of set deforestation. . RENEW your Commitment to lead the way to save the planet